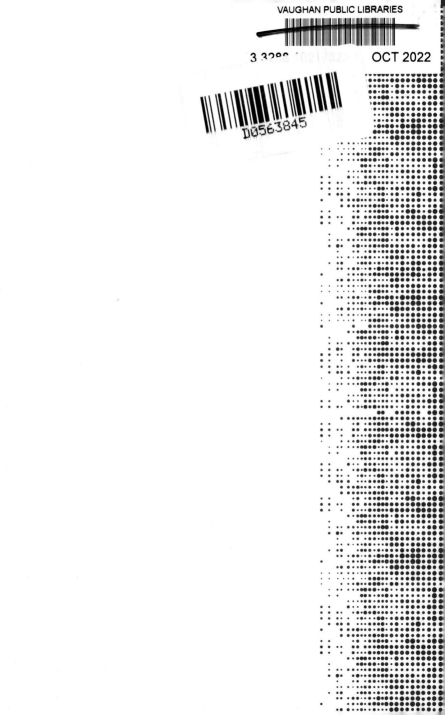

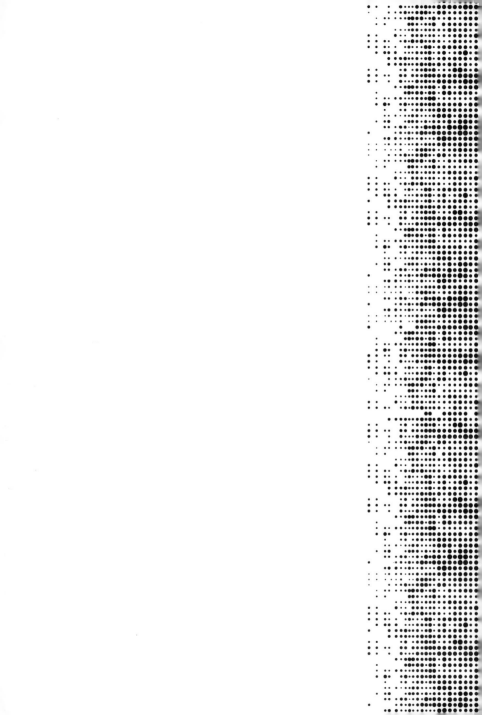

Copyright © 2022 Blue Star Press
Published by Blue Star Press
PO Box 8835, Bend, OR 97708
contact@bluestarpress.com | www.bluestarpress.com

Written by John-Paul Garrigues
Design and illustrations by Chris Ramirez
Photography by John-Paul Garrigues

ISBN: 9781950968800

Canon® is a registered trademark of Canon U.S.A.

Epson® is a wholly-owned subsidiary of Seiko Epson Corporation. EPSON, the EPSON logo, and the EPSON Exceed Your Vision logo are registered trademarks of Seiko Epson Corporation ("SEC"), registered in the U.S. and other countries.

iPhone® is a trademark of Apple, Inc., registered in the U.S. and other countries. ImageWriter® is a trademark of Apple, Inc. registered in the U.S. and other countires. Apple® is a trademark of Apple, Inc. registered in the U.S. and other countries.

HP® and Hewlett-Packard® are registered trademarks of Hewlett-Packard Company.

All other referenced trademarks and registered trademarks are the property of their respective companies.

Sh*tty Printers is entirely satire, and is to be considered a work of fiction for entertainment.

Printed in Colombia.

10 9 8 7 6 5 4 3 2 1

Sh*tty Printers

Foreword

While writing the foreword to the very book that you hold in your hands, something preposterous happened:

A picture, very likely *not* painted by Renaissance master Leonardo Da Vinci, sold at auction for 450 million US dollars, thus jeopardizing my entire plan for this piece. Because up until this moment, the printer—specifically, in my opinion, the at-home inkjet printer—was the single object most emblematic of man's folly, of his casual, unwitting cruelty to himself and to his fellow man.

In 2011, the gross domestic product of the island nation Tonga was 439 million USD.

Consider the printer.

It is a machine that delivers on exactly the opposite of the promise of the Information Age. And it is therefore the truest manifestation of the era.

From the very start, the promise of the personal computing revolution was two-fold: all the world's knowledge at one's fingertips, and streamlined communication for both personal and professional reasons. And as soon as there were personal computers, there were printers, fucking things all up, taking two giant leaps back for every one step forward we took into the future. Putting what computers worked so hard to digitize back, needlessly, into the physical world.

"So you mean to tell me," you might have asked the clerk at CompUSA in 1999, "that now I can fill out this medical form entirely, as you say, 'online?'"

"Oh, no," he'd have told you, sagely shaking his head. "Digitally signing things won't be hot in these streets for at least another
10 years."

"So I—"

"What you can do now, here, in 1999, is you can print out a copy of the medical form from your email, sign it, and send it back to them. But in order to do that, you'll need one of these hellish 3-in-1 machines." He would have turned then on the heel of his all-black sneaker, and, with a flourish, gestured—as though revealing from behind an imaginary curtain—at several of these printers with the flatbed scanners bolted to the top and made by companies like Lexmark, Canon, and Epson.

"I understand that it prints and scans. What's the third function?"

"Copying or faxing, probably, but I can't be certain. No one has ever attempted either."

It didn't matter then and it doesn't matter now, because the fact remains that, on top of a filing cabinet, or in the side table in some corner of literally every home office in the world, there squats one of these insane devices. Soldiering away. Taking virtual content from computers, and making documents that resemble that content using ink that squirts out of little tanks and paper and dozens of little gears and servos all designed with planned obsolescence in mind. Little skeuomorph factories, immovable objects, always ignored, and in constant need of maintenance.

You can print from an iPhone®. It's the dumbest thing.

This book is called **Sh*tty Printers**. From the title, and this foreword so far, you might infer that the author and myself believe all printers to be shitty. Such is not the case. Those crazy large-scale plotter printers that they make blueprints with? Those things are fantastic. Truly lovely, honorable printers were used in the production of this very tome, and we owe them all a debt of gratitude. Because, honestly.

The author of this book, photographer and humorist J.P. Garrigues, has a weird, uncanny knack for zeroing in on exactly what's bothering you, before you've even noticed it. "Do you hear that," J.P. will ask you, walking into a room, pointing out the barely perceptible high whine of a television. And then you can't hear anything else.

Perhaps you were unaware of the combined years of anguish and outrage printers have caused you. Flip through this book and you'll find one, for sure, that has personally offended you in some way, and you'll remember the three consecutive trips to the Circuit City™ it took to get the appropriate ink cartridges. Or the hours spent on hold, waiting for customer service. Or the time the whole thing stopped working altogether when it was really important that it functioned correctly. Remember these things, revel in them, and thank J.P. for holding a mirror to all the trauma you thought you had forgotten.

— Bo Fahs

Bo Fahs is a writer and the co-host of the podcast Tele-Friends. He lives in Austin, Texas.

Introduction

Nature has produced creatures that hiss, rattle, snarl, or emit toxic substances from glands that shouldn't otherwise exist in a magnanimous-diety-created world. Yet these abilities evolved, discouraging other animals, especially humans, from approaching or making any sort of contact.

Printers, whether through design or spontaneous sentience, have developed similar warnings—warnings that we as a stubborn species blithely ignore. They scream at plane-like decibels, interfering with our hearing and basic human interactions. They emit sweet-smelling ozone, inviting us to inhale the tissue-destroying free radicals. They trigger cortisol to be released regularly, shortening our lifespans. Yet, we continue to invite them in our home-offices, quaint cubicles, and reception desks worldwide.

Printers are the self-sabotaging gatekeepers between the digital world and the physical world. Their purpose is to translate and act as a medium between our fruitful, corporeal existence and the upside-down. We are but humble primates, routinely sacrificing sheets of cellulose as offerings to the beige obelisks before us. And while the trees remain giving, the printers continue to taketh away.

Whether or not their perceived hostility is intentional is up for debate. They could be Cenobithian harbingers, here to deliver both pleasure and pain; it's all the same to them.

Or, maybe we're the problem. What we interpret as devils hellbent on tearing our life away, may actually be angels freeing us from an ugly and unfamiliar world. Perhaps humanity has long since passed and printers are merely the black-cast shadows of a deteriorating dreamworld.

What is clear, however, is that we created this mess. We did this to ourselves, and this book is a chronicle of but a small sample of our own masochism and the shittiness within.

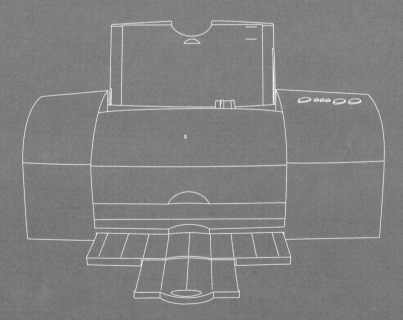

HP®
ThinkJet
2225A

HP ThinkJet 2225A

The piece of shit that started it all.

The HP Thinkjet was the first widely available consumer inkjet printer, and the start of the decline of human civilization. Decades from now, documentaries will feature slowly panning photo montages of Pol Pot, Khrushchev and the ThinkJet 2225A as symbols of man's hubris and inherent fallibility.

HP is an ominously apt name for the company that unleashed such a nightmarish object into this world. Hewlett Packard. HP. H.P. Lovecraft. It's all there: WAKE UP SHEEPLE. Only Lovecraft and a huge multinational conglomerate could devise such ghoulish, sadistic creations.

STRENGTHS:
Small size. Can transcend the physical realm,
entering our deepest and most intimate thoughts.

WEAKNESSES:
Unappreciated during its lifetime.

CHARISMA:
Nyarlathotep.

RATING:
5 / 10 Cthulus

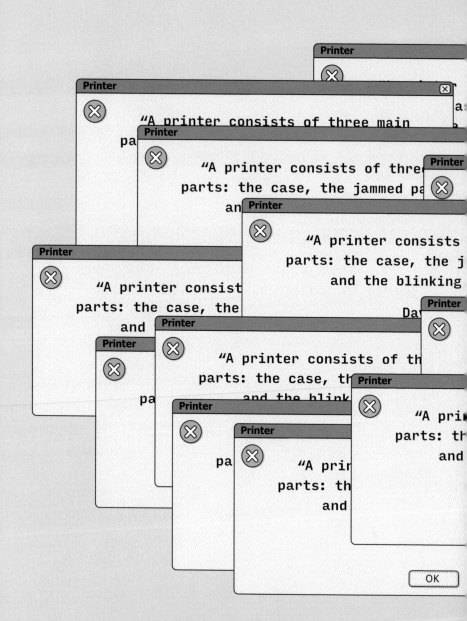

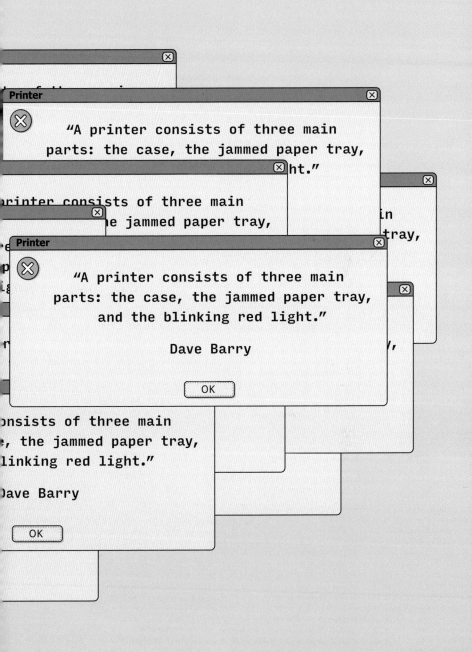

Printer

"A printer consists of three main
parts: the case, the jammed paper tray,
and the blinking red light."

Dave Barry

OK

LEXMARK® Z22

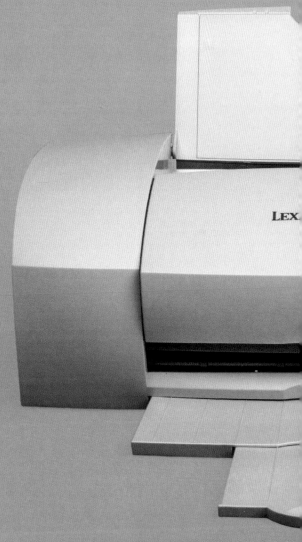

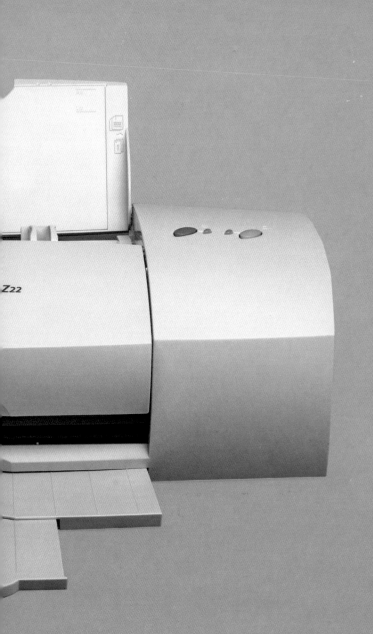

LEXMARK Z22

The Sophie's Choice of print technology.

Consisting of only two buttons, your options for customization are either shitty teal or shitty gray. Owning a Lexmark Z22 was like spontaneously growing an all-seeing, omniscient third eye—one with debilitating cataracts. The initial bliss of owning a Z family, full-color inkjet, with questionably-futuristic turquoise buttons and a nouveau-curved form factor, subsided as the machine promptly showed its true colors (mainly shitty-gray and shitty-teal).

The first sign of failure was the streaks that invariably appeared on only the most important printouts. Somehow, Lexmark must have designed a proprietary chip to differentiate and destroy important documents like résumés and book reports, and yet preserve non-essential printouts. Sometimes the streaks were purely in the cyan. Sometimes they were in the black. Sometimes they appeared as digital errors, but they always felt like organic middle fingers that taunted you, and goaded you to try a different manufacturer.

STRENGTHS:
All of your print-making decisions have already been made for you.

WEAKNESSES:
Lack of customization.

CHARISMA:
Tacky teal tints.

RATING:
8 / 10 regrets ● ● ● ● ● ● ● ● ○ ○

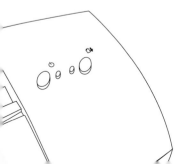

BILL KELLER:

"Every technology, including the
 printing press, comes at some price."

(Show Printer)   (OK)

Hard Drive

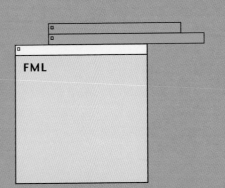

FML

BLUE CHIP
D12/10

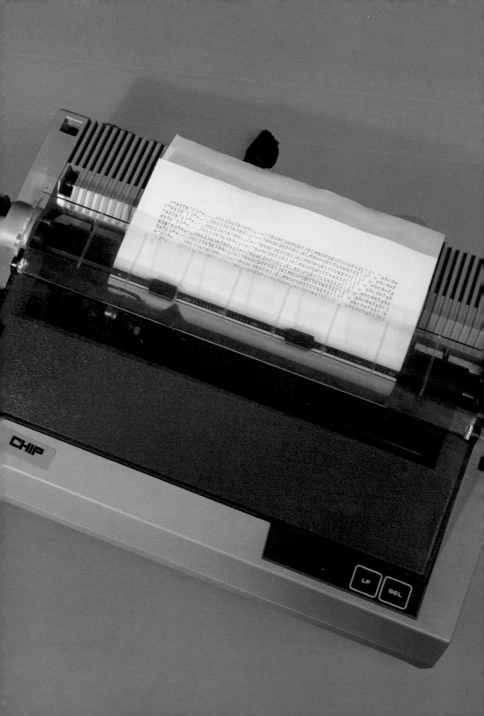

BLUE CHIP D12/10

The term 'Blue Chip' is a phrase that typically connotes high value or high quality.

D12 typically refers to the Dirty Dozen—a rap super group from Detroit. The Blue Chip D12/10 is none of these things. What this naming convention or brand name is trying to convey is unclear, and this printer is just a complete mess overall. The manufacturer couldn't even affix the sticker on straight, let alone make the printer work.

The biggest drawback of the D12/10— aside from its ambiguous name—is the excruciating pace at which it shits out pages. At an advertised yet rarely realized twelve characters per second, you'd be better off conveying your message by learning Morse code, inking it out by hand, and then hot-gluing your prose to the head of a sickly pigeon.

STRENGTHS:
Simple button layout.

WEAKNESSES:
Sloppy execution, agonizingly slow.

CHARISMA:
Eminem and Bizzare's shroom-induced love child.

RATING:
2 / 10 Mom's spaghettis

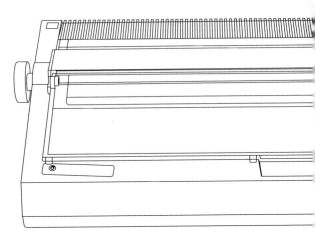

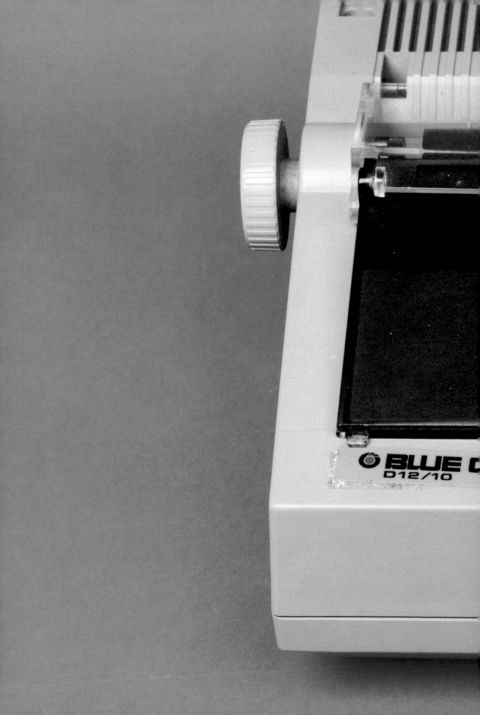

"The world has been printing books for 450 years, and yet gunpowder still has a wider circulation. Never mind! Printer's ink is the greater explosive: it will win."

Christopher Morley

"The world has been printing books for 450 years, and yet gunpowder still has a wider circulation. Never mind! Printer's ink is the greater explosive: it will win."

"The world has been printing books for 450 years, and yet gunpowder still has a wider circulation. Never mind! Printer's ink is the greater explosive: it will win."

Christopher Morley

Christopher Morley

HP
Deskjet® 3050A

HP Deskjet 3050A

The 3050A e-All-in-One scanner/printer is a testament to mankind's contribution to entropy.

Accolades to the designers for going straight to the ultimate state of this piece: rust. HP was able to combine the beautiful aesthetic shade of corroded iron oxide with the undying longevity of non-degradable plastic.

Humans may not endure, but the shittiness of this printer will. Despite looking decrepit from the get-go, this totem to humanity's faults will be around long after mankind has departed this mortal coil.

STRENGTHS:
Longevity.

WEAKNESSES:
Initial quality.

CHARISMA:
Tetanus shots in all the wrong places.

RATING:
10 / 10 Ozymandiases

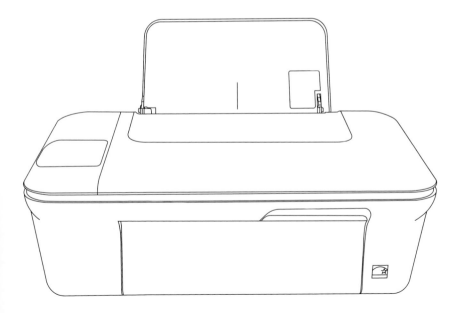

Look on my works,
ye Mighty, and despair!
Nothing beside remains.
Round the decay
Of that colossal wreck,
boundless and bare
The lone and level sands
stretch far away.

— Ozymandias

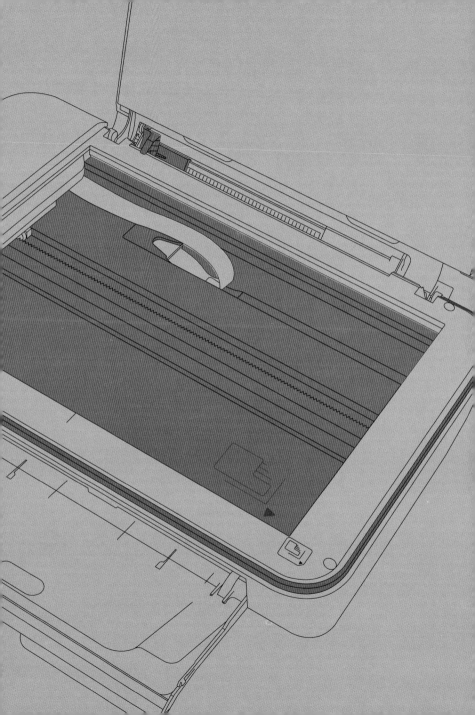

Brother
HR-15XL

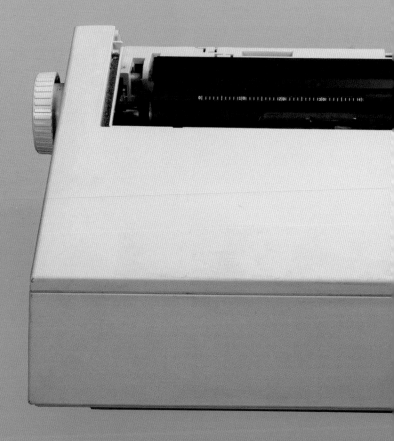

Brother HR-15XL

Oh brother, what have we become?

A vapid husk of our former selves. Anyway, this printer:

TofLfSel. The menu of buttons on the facade of the printer reads like a New York City borough portmanteau. But rather than chauffeuring you to some super sexy fashion week, this ride drags you through the gutter grease on Canal Street while some elderly crone yells "Vuitton?" at you from the shadows. The dark elbow of Doyers Street is the only salvation until the Tong find you and force you to print test page after test page.

Some say that late at night, the HR-15XL will appear near Manhattan & Nassau, sentenced to eternally ride the ghost line.

STRENGTHS:
Completely resistant to subway poo.

WEAKNESSES:
Can't elegantly fit through the turnstile.

CHARISMA:
Harlem Boppers' purple vests.

RATING:
7 / 10 Subway stabs

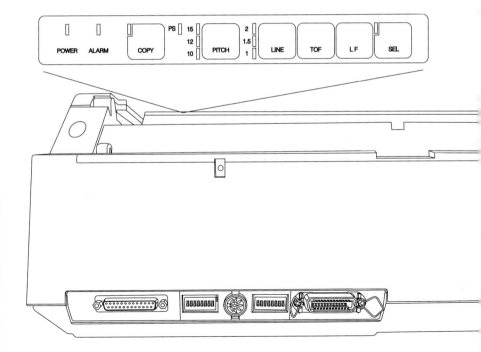

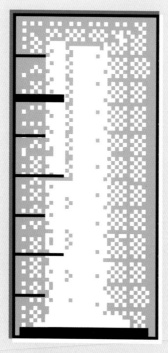

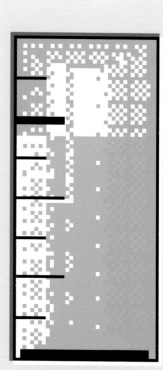

Black:
0%

Cyan:
69%

"The printing press is either the greatest
blessing or the greatest curse of modern

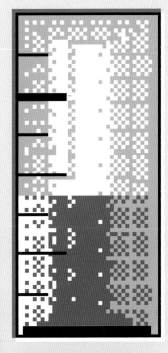 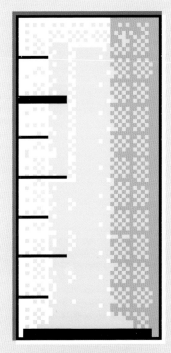

Magenta:
37%

Yellow:
98%

* Estimate only.
  Actual ink levels may vary.

Apple® ImageWriter™ II

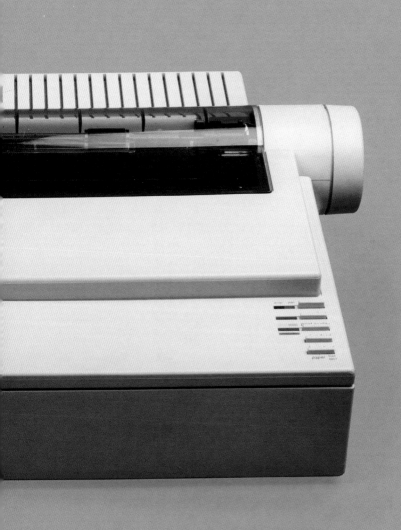

Apple ImageWriter II

The Apple ImageWriter— so beloved by tinnitus patients that Apple decided to release a second version.

The ImageWriter II shook like none other; it was the purely distilled manifestation of involuntary tremors on bath salts. And to top it off, it also screeched like no other. Before fracking was a thing, earthquakes were triggered by this undulating banshee's unyielding yawps. The ImageWriter may have been trying to communicate and convey the eternal pain of its own existence. Or perhaps it was just really shitty.

STRENGTHS:
Incredibly dense. Unimaginably heavy for its size.
Actually reliable.

WEAKNESSES:
Can't contain itself.

CHARISMA:
Boogaloo Shrimp Chambers.

RATING:
7 / 10 Krumpers

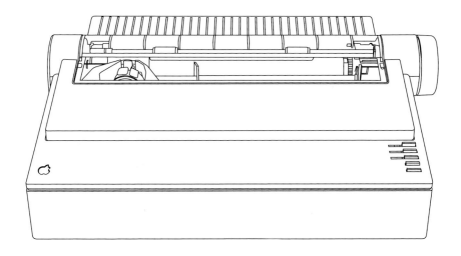

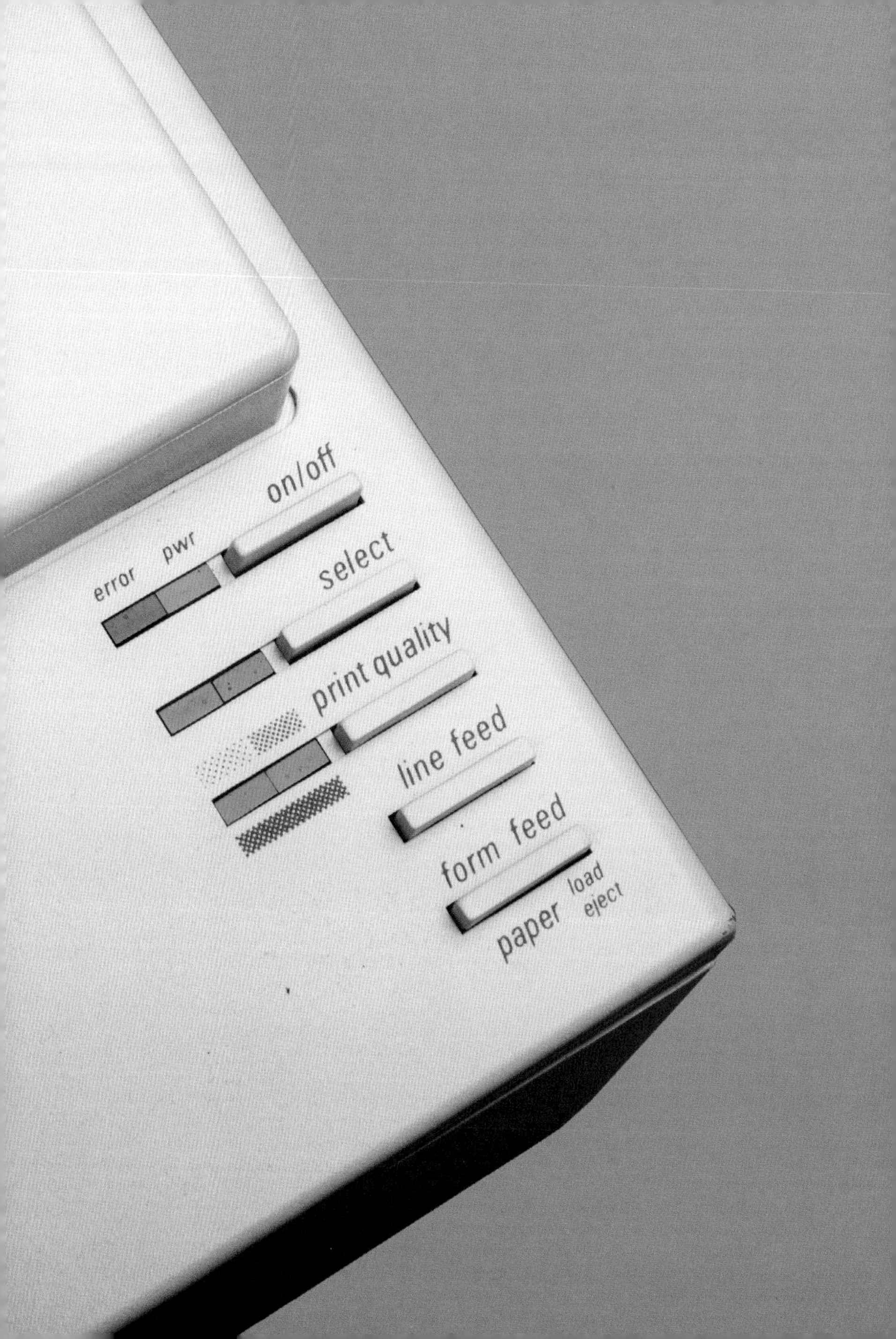
on/off
error
pwr
select
print quality
line feed
form feed
paper
load
eject

STAR
NX-1000C

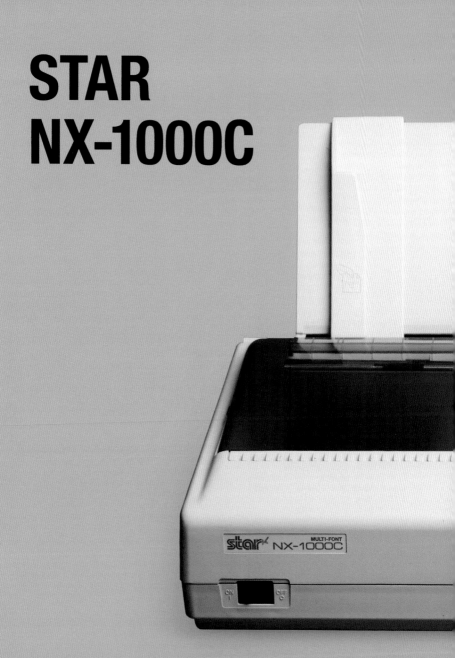

STAR NX-1000C

Star printers are actually beautifully designed.

Compact, color-coordinated, forward-thinking minimalism without any bullshit. They just work. Complete anomalies in a sea of mediocrity. So why even include these works of art in this compendium?

The ink in these things smells like vomit.

There's no way around it. If you own a Star printer and want to print your great American novel in that mess you call your home office, then you might as well just hold your finger down your throat until you vomit gingerly into your Glade® Plugins®. Every piece of literature printed on the NX-1000C becomes a polluted serviette saturated with the nebulized refuse of the human gullet.

STRENGTHS:
YOU WERE SO CLOSE, STAR.

WEAKNESSES:
Smells like upchucked migas floating in potato water.

CHARISMA:
Super-sexy Kobayashi with a stomach virus.

RATING:
4 / 10 Ralphs

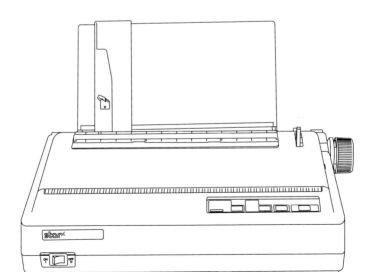

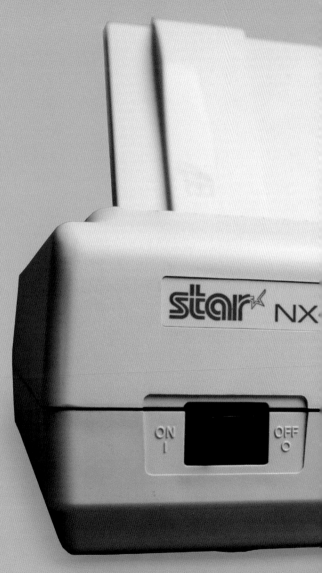

So fucking close.

STAR
SL-10C

STAR SL-10C

Break out the Vicks VapoRub.

If you acquire enough rub, you should be able to tolerate the fact that your office smells like a third-rate frat house after a particularly vicious hazing sesh. And if so, Star printers are at your service. This beautifully minimalistic printer is not-fuck-withable. Credit is due on the solid construction, minimalist design, and versatility. Designed to work on old Commodores, the Star SL-10C was perfect for old-school modders and trackers to print out their acid jazz compositions.

Where the SL-10C falters, however—and the reason it is deemed a shitty printer—is the completely ridiculous packaging surrounding this majestic icon of design. Seemingly conceived by a poor man's Robert Crumb, the whimsical instructions on the back of the box were the antithesis of a cheap thrill. Rather than help them, the hokey saccharine illustrations worked more to mock the owner. In fact, this may be what set off Rob's brother Charles Crumb, causing him to abandon printers altogether and fill up stacks of hand-written notebooks in the attic.

STRENGTHS:
It's a Star printer, homie.

WEAKNESSES:
Packaging and instructions more than likely created by a socially-inept misogynistic racist.

CHARISMA:
Cellulite thighs.

RATING:
9 / 10 Giant asses

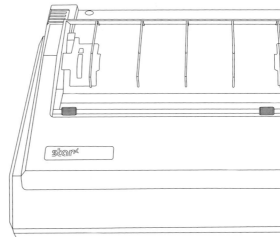

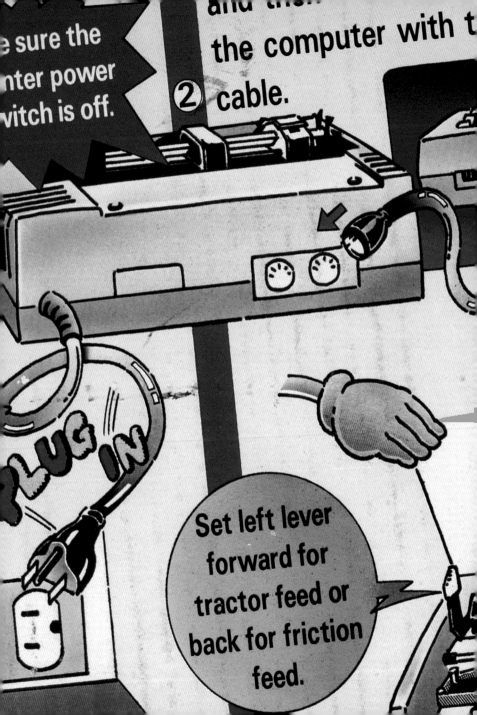

ncluded

CONNECT

OR

Star SL-10C
box art

Put

HP
Officejet® J4550

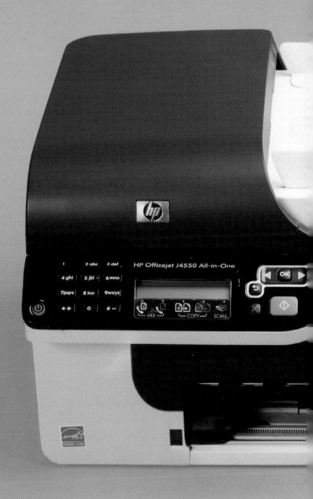

HP Officejet J4550

In the early 2000s, long before they got all Skrillexy with their logo, HP developed a brilliant strategy of just completely fucking phoning it in.

From using some spooky Halloween font in year-round advertising campaigns to selling their competitors' products (ex: Apple iPods painted shitty blue), HP couldn't disguise the fact that they just did not give a fuck. HP did not give two shits about the customers, their employees, quality control, their flagship products, or even making money for that matter. The black-hole of a deal known as "the Compaq acquisition" was the likely culprit, draining brain-power from the corporation and spiraling it towards complete mediocrity. All of that core inadequacy is distilled into the OfficeJet J4550. This thing jammed and not in a bro-step, finger-in-the-air kind of way. Prone to paper, carriage, and ink jams, the J4550 was like all of the boring throwaway parts of shitty dubstep, with none of the angry robot stuff.

STRENGTHS:
Committed to the blue thing.

WEAKNESSES:
Wasn't an Apple® product.

CHARISMA:
A gaggle of bros yelling "ohhhhh shit".

RATING:
10 / 10 Douche Goggles

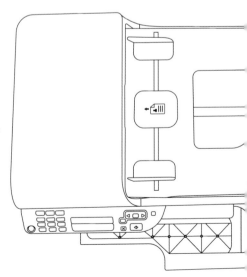

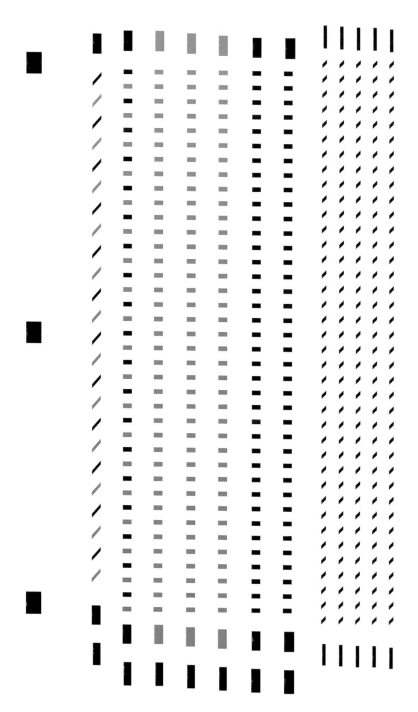

IMPORTANT!
You must scan this alignment page for best print quality.

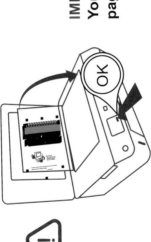

Canon®
BJC-85

Canon BJC-85

The '90s were the absolute apogee of human development.

Curiosity was still a fundamental characteristic of what differentiated us from the feces-flinging primates we pitied so condescendingly. It was a time where the internet was richly rewarding, but not completely ingrained into our daily lives. A time where steampunk was primarily relegated to Will Smith's *Wild Wild West* rather than displayed by self-loathing, polyamorous, middle-aged white people. Somewhere between TRL and The Ginger (aka IT, aka Segway), everything went to shit. Human curiosity was replaced by machine learning, and our ability to better ourselves gave way to curated timelines fed to us by devices that learned their social graces from otaku.

The BJC-85 is the representative inflection point of this change. The younger brother of the BJC 4000 itself is great—it's incredibly light, well designed, ultra-portable, and has parallel, USB, and infrared interfaces. Not only that, but you can replace the print cartridge with a scanner cartridge transforming the whole device into a portable scanner. However, the BJC-85 never really caught on, and cheap imitators such as the Z22 became ubiquitous.

STRENGTHS:
Ultra-portable, convertible, sounds like an Autechre B-side.

WEAKNESSES:
Color bleeding, INSANELY slow scanning, potentially responsible for human decline.

CHARISMA:
Carson Daly's pinky nail.

RATING:
9 / 10 Lou Pearlmans

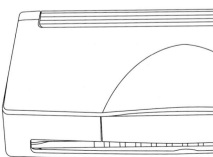

LETTER
LETTER
LETTER
LETTER
LETTER
LETTER
LETTER
LETTER

Epson®
Apex 80

Epson Apex 80

"Everything is a commodity, including people."

In 1987, retired actor, decaying dream-boat, and American president Ronald Reagan solicited Mikhail Gorbachev to tear down that wall. Two years later, the world cheered as the Cold War evaporated and the not-so-slow march of capitalism overtook the last vestige of western communism. What wasn't well known at the time was that a KGB officer on the Eastern side was in mourning, quietly seething, and he would resent this moment for decades. That officer's name was Vladimir Putin, and this specific event would lead to targeted subterfuge and the eventual vaporization of American ideals.

Perhaps not coincidentally, the Epson Apex 80 was also released in 1987. Brutalist by design, the printer's facade represented a patriarchal, fascist approach to consumer technology. You did what it fucking told you to do! You may have spent your last few paychecks bringing this appliance into your warm bungalow, but this beige beast held the leash.

In a classic Vladislav Surkov move, Epson not-so-subtly sprinkled hints of capitalism throughout. The unadorned exterior was hatcheted with a haphazardly graffitied product emblem. The menu system was set up as if to pay homage to a caste system, and typography options such as "Elite" and "DOUBLE STRIKE" graciously trickled down upon unappreciative plebes.

Epson Apex 80

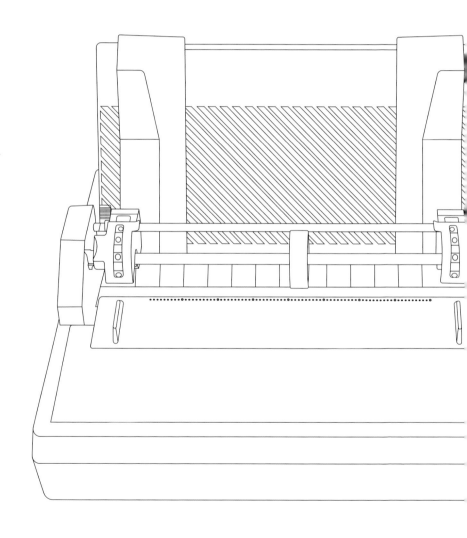

STRENGTHS:
Subterfuge.

WEAKNESSES:
Vodka.

CHARISMA:
Riding a bear shirtless.

RATING:
6 / 10 Comrads

Recycle Bin

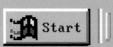

Warning ✖

 "PC load letter; what the fuck does that mean?"

[Mike Bolton]

🖨 9:59 PM

LEXMARK
X1240

LEXMARK X1240

"They hate us, you know...the humans. They'll stop at nothing."

Rumor has it the X1240 was a special project jointly developed by NASA and Kubrick during his research phase for the production of *A.I.*. By this time Lexmark really started to get their act together and graciously offered more than two tactile options for their bevy of buttons. Kubrick collected boxes and boxes of product design references in an attempt to imitate the inherent beauty of well-designed machines, creations of love and grace.

However, even a master like Kubrick can eventually falter, and he came to realize the folly of a partnership with Lexmark. Kubrick scrapped his endeavors and decided to go a different direction: give *A.I.* to Spielberg and put Tom Cruise in a movie about fucking. It worked. This printer, however, did not.

Supertoys may last all summer, but the Lexmark X1240 existence was fleeting, fledgling, and flat.

STRENGTHS:
Clean design, future-proof.

WEAKNESSES:
It's a Lexmark.

CHARISMA:
Chris Cunningham's love assembly bots.

RATING:
1 / 10 Haley Joel Osments

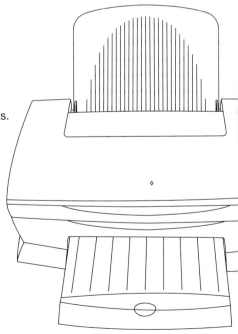

Brother
HL-L2380DW

Brother HL-L2380DW

Aside from the wireless setup process, this printer is actually pretty good.

The scanner works, drivers are relatively seamless, it prints as advertised, and is reasonably priced. Overall, not bad. It is, however, the only monochromatic laser printer present in this book and is representative of some of the worst aspects of laser printing. Laser printers are high-voltage, energy demanding, toxin-spewing, nigh explosive death machines. Although to date there have been very few incidents of printers being hacked and set afire, it's only a matter of time before these death machines rise up and exact their will upon their creators.

Go to sleep now knowing that if you wake up tomorrow, this machine is ready to go to town.

STRENGTHS:
Deceptively acquiescent.

WEAKNESSES:
Wireless printing setup is a hint of things to come.

CHARISMA:
"I'd fuck me."

RATING:
5 / 10 Dahmers

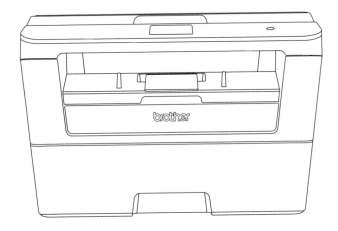

MEMO

MEMO